PRACTICE
Workbook

Grade K

Harcourt
SCHOOL PUBLISHERS

Visit *The Learning Site!*
www.harcourtschool.com

CALIFORNIA HSP Math

ISBN 13: 978-0-15-356908-1

ISBN 10: 0-15-356908-5

13 14 15 1421 16 15 14 13 12 11

4500335828

Contents

UNIT 3

Chapter 5: Graphing

Chapter 6: Geometry

UNIT 4

Chapter 7: Numbers 11 to 20

Chapter 8: Numbers 21 to 30

UNIT 5

UNIT 6

Name_____

Alike and Different

DIRECTIONS 1-4. Mark an X on the object that is different.

Name_____

Hands On: Sort and Classify by Color

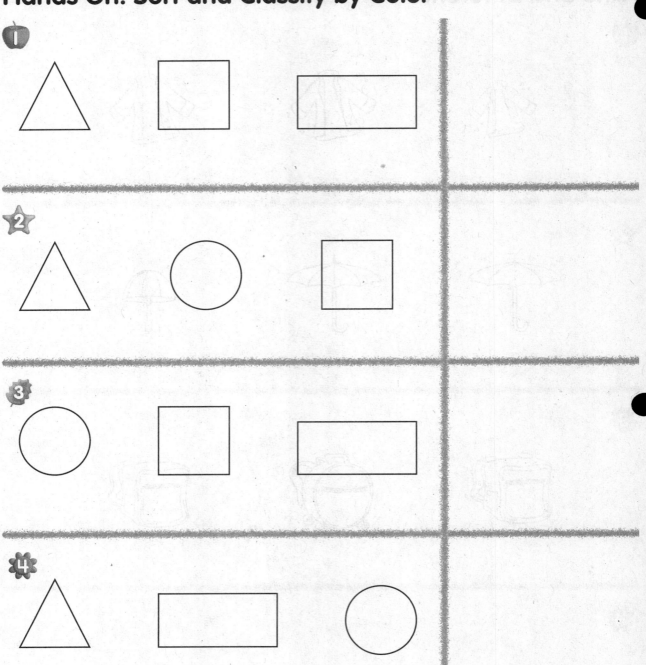

DIRECTIONS 1. Use red to color the figures at the beginning of the row. Draw and color one more figure that would belong in that group. 2. Use blue to color the figures at the beginning of the row. Draw and color one more figure that would belong in that group. 3. Use yellow to color the figures at the beginning of the row. Draw and color one more figure that would belong in that group. 4. Use green to color the figures at the beginning of the row. Draw and color one more figure that would belong in that group.

Practice

Name_____

Hands On: Sort and Classify by Size

DIRECTIONS 1–4. **Mark an X on the object that does not belong.**

Practice

Name_____

Hands On: Sort and Classify by Shape

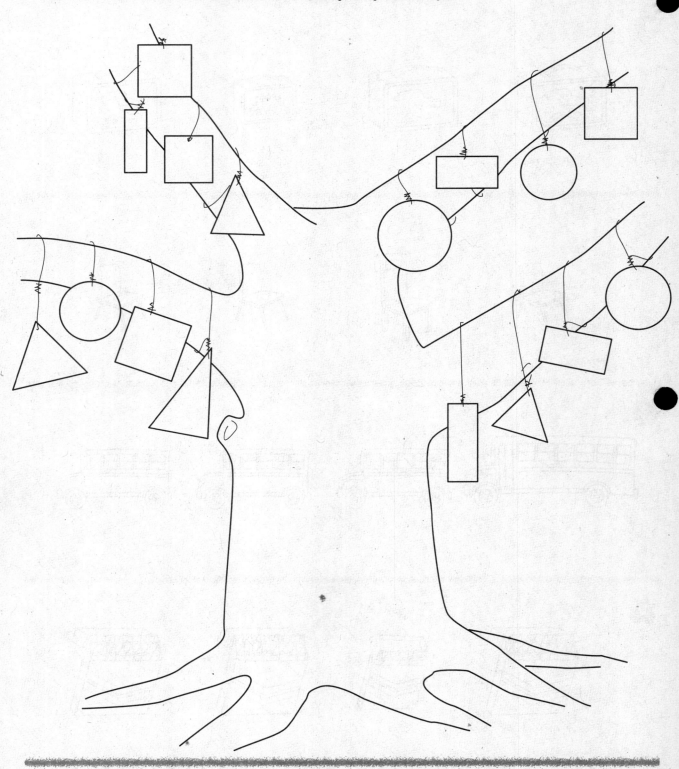

DIRECTIONS Circle the two figures that are alike on each branch. Draw and color those figures on the trunk of the tree.

Practice

Problem Solving Workshop
Skill • Use Visual Thinking

DIRECTIONS 1–4. Color the first figure in the row. Color all the other figures to match the color of the figure at the beginning of the row. Circle one figure that matches in more than one way. Tell how the figure matches.

Practice

Identify Sorting Rules

1

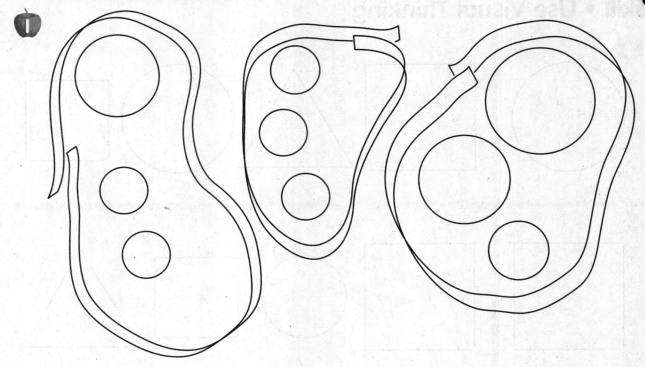

2

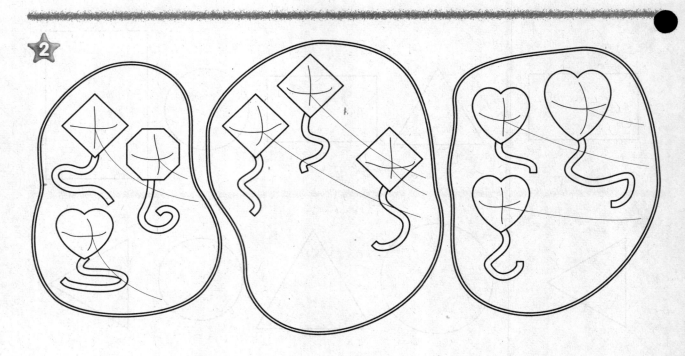

DIRECTIONS   1–2. Draw one more figure in the group that is sorted by shape and size.

Practice

Name_____

Name _____

Name _____

Problem Solving Workshop
Strategy • Use Logical Reasoning

Lesson 1.7

DIRECTIONS 1–4. Three of the objects are alike. Mark an X on the one that does not belong. Tell why it does not belong.

PW7

Practice

Hands On: Sort to Make a Graph

1

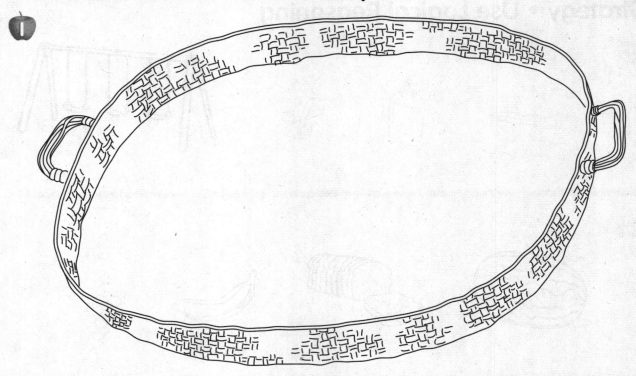

2

Blue and Red Cubes

DIRECTIONS

1. Place a handful of red and blue cubes on the basket. Sort the cubes by color.
2. Move the blue cubes to the top row on the graph. Use blue to color the cube at the beginning of the top row. Move the red cubes to the bottom row on the graph. Use red to color the cube at the beginning of the bottom row. Then draw and color the cubes on the graph.

Practice

Name_____

Hands On: One-to-One Correspondence

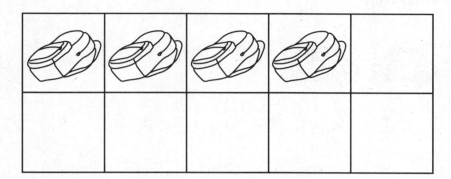

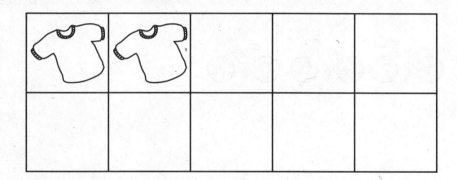

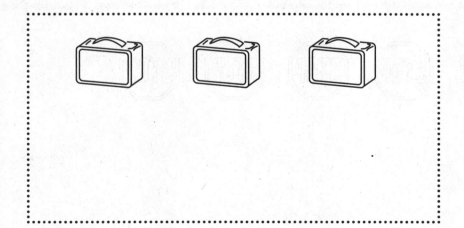

DIRECTIONS 1–2. Place a color tile below each object to show an equal set. Draw and color each color tile. 3. Draw a juice box below each lunch box to show equal sets.

Practice

Problem Solving Workshop
Skill • Use a Model

DIRECTIONS   I–3. Use cubes to show a set of fewer objects. Draw the cubes.

Practice

Hands On: Model, Read, and Write 1, 2, 3, 4

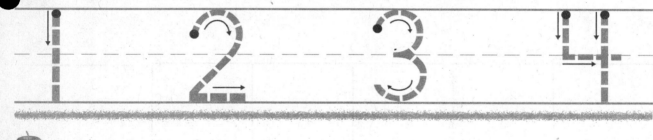

_____

_____

_____

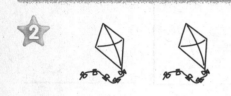

_____

- - - - - - - - - - -

_____

_____

- - - - - - - - - - -

_____

DIRECTIONS 1–4. Place a cube on each object or animal in the set as you count. Draw the cubes. Write the number.

Hands On: Model on a Five Frame

 1

2

3

4

DIRECTIONS 1–2. Use color tiles to model three. Draw the tiles. Write the number.
2. Use color tiles to model two. Draw the tiles. Write the number.
3. Use color tiles to model four. Draw the tiles. Write the number.
4. Use color tiles to model five. Draw the tiles. Write the number.

Practice

Read and Write 5

5 5 5 5 5

_____
- - - - - - - -
_____

_____
- - - - - - - -
_____

_____
- - - - - - - -
_____

DIRECTIONS 1. Trace the number 5.
2–5. Write the number that tells how many in the set.

Read and Write 0

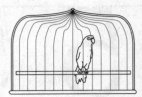

zero

_ _ _ _ _ _

_____

one

_ _ _ _ _ _

_____

 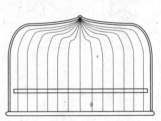

zero

_ _ _ _ _ _

_____

two

_ _ _ _ _ _

_____

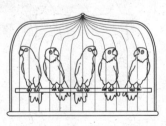

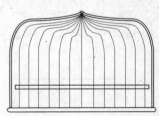

five

_____

_ _ _ _ _ _

zero

_____

_ _ _ _ _ _

DIRECTIONS 1–6. Write the number that shows how many birds are in the cage.

Write Numbers to 5

 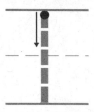

DIRECTIONS 1–6. Trace the number that shows how many objects are in the set. Then write the number.

Practice

© Harcourt

Problem Solving Workshop
Strategy • Make a Model

DIRECTIONS   **I.** Trace the number of links.
2. Make a chain that has I link more than 2. Draw the chain. Write the number.
3. Make a chain that has 2 links more than 2. Draw the chain. Write the number.

Name_____

Hands On: Order Numbers to 5

1

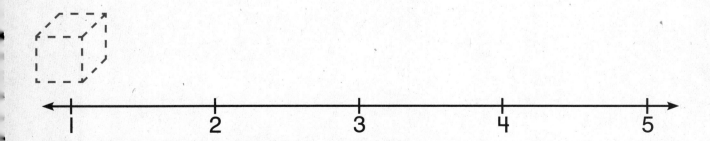

DIRECTIONS **1.** Make a cube train to model each number. Draw the cube trains.

Top right:

Lesson 2.9

Practice

Hands On: Identify Patterns

 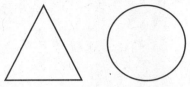

DIRECTIONS 1–3. **Step 1.** Read the pattern. **Step 2.** Place figures to identify the pattern. **Step 3.** Trace and color the pattern.

Describe Patterns

 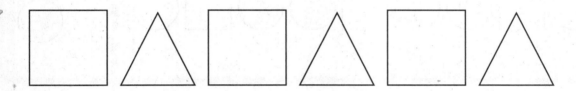

 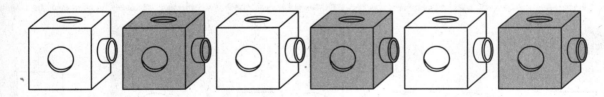

DIRECTIONS I. Use bear counters to copy the pattern. Describe the pattern. Draw and color the pattern. 2. Use figures to copy the pattern. Describe the pattern. Draw and color the pattern. 3. Use cubes to copy the pattern. Describe the pattern. Draw and color the pattern.

Practice

Hands On: Identify and Extend Patterns

 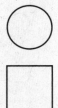

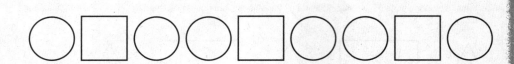

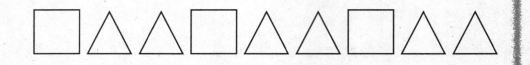 

 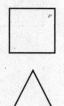

DIRECTIONS   I– 4. Use figures to identify the pattern. Circle the figure that most likely comes next in the pattern.

Problem Solving Workshop
Strategy • Find a Pattern

2

3

DIRECTIONS 1–3. Look at the pattern. Write the number of spots below each ladybug. Circle the part that repeats again and again.

Hands On: Identify and Create a Pattern

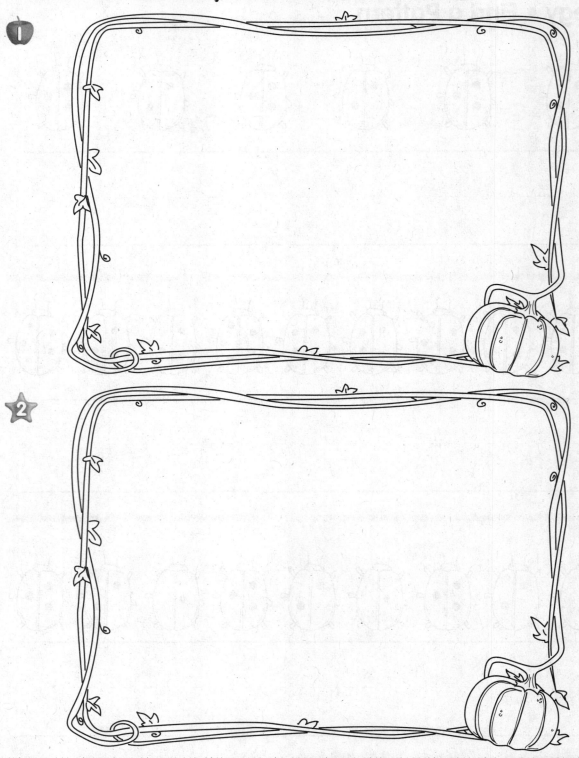

1

2

DIRECTIONS 1–2. Use Attribute Links to make a color, shape, or size pattern. Draw and color your pattern. Tell someone about your pattern.

Problem Solving Workshop
Skill • Use a Pattern

1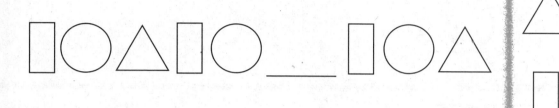

2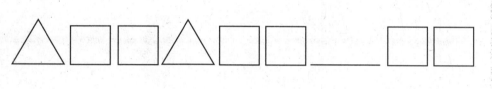

3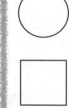

4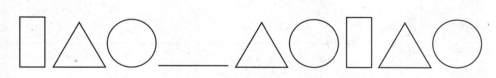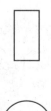

DIRECTIONS   1–5. Circle the part that repeats in each pattern. Circle the figure that is missing.

Hands On: Model, Read, and Write 6 and 7

 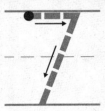

DIRECTIONS 1–2. Use counters to model the number. Draw the counters. Say each number as you trace it. 3–4. Count the objects in the set. Trace the number. Write the number.

Practice

Hands On: Model, Read, and Write 8 and 9

 1

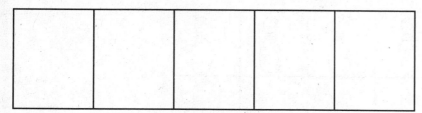 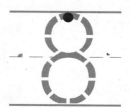

2

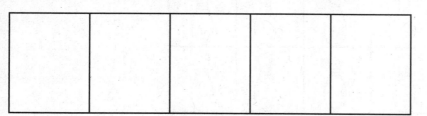 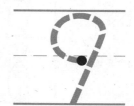

3

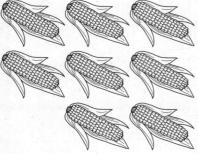

_____

- - - - - - - -

_____

4

_____

- - - - - - - -

_____

DIRECTIONS 1–2. Use cubes to model the number. Draw the cubes. Trace the number. 3–4. How many objects are in the set? Write the number.

Practice

Hands On: Model on a Ten Frame

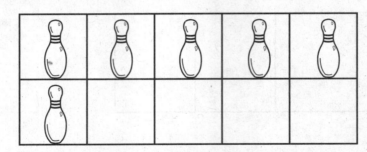

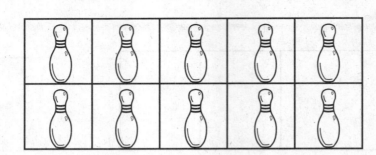

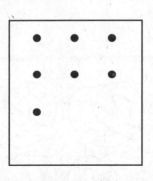 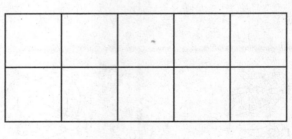

 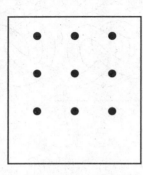 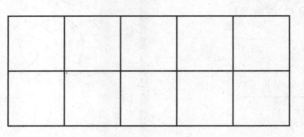

DIRECTIONS 1–2. How many bowling pins? Write the number.
3–4. How many dots? Place counters in the ten frame to model that number.
Draw the counters. Write the number.

Name____

Read and Write 10

1

10 10

2

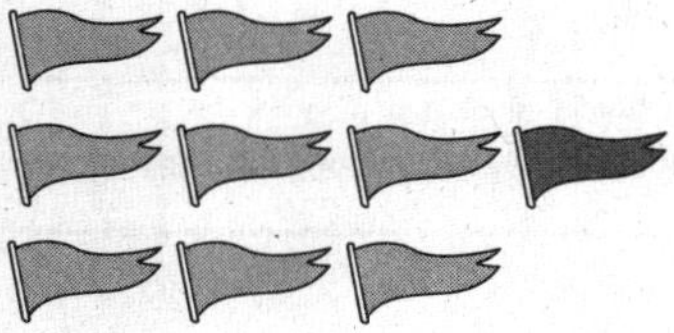

3

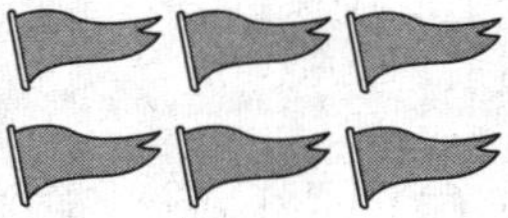

4

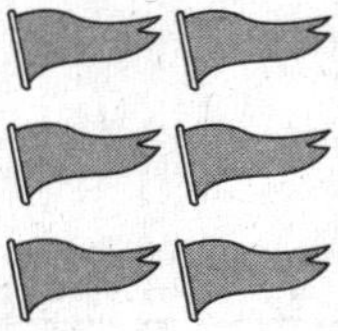

5

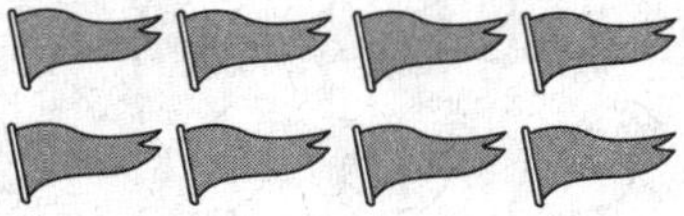

6

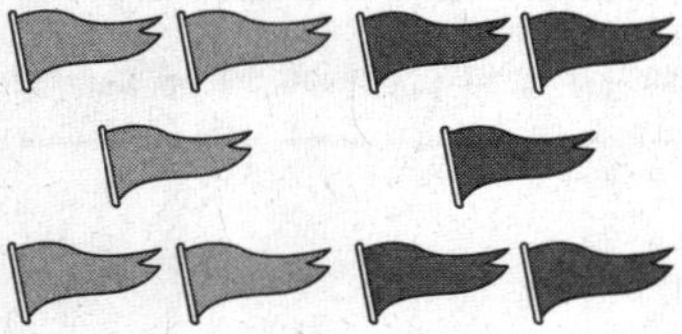

7

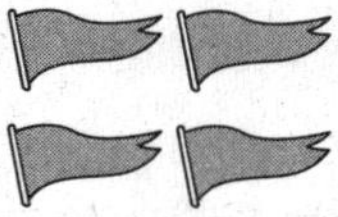

DIRECTIONS **1. Trace and write the number 10.**
2–7. How many flags? Write the number. Circle the sets that have ten flags.

Name_____

Write Numbers to 10

 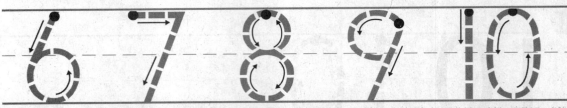

_____

_____

_____

_____

_____

_____

_____

_____

_____

_____

DIRECTIONS 1. **Say the number as you trace it. 2–6. How many birds? Write the number.**

Practice

Name_____

Lesson 4.6

Hands On: Compare Sets to 10

DIRECTIONS 1–3. Place a red cube on each horse. Place a blue cube on each child. Join the cubes for each set. Compare the cube trains. Write how many in each cube train. Circle the set that has more.

PW29

Practice

Problem Solving Workshop
Strategy • Make a Model

1

_____   _____

- - - - - - - -   _____

_____   _____

2

_____   _____

- - - - - - - -   - - - - - - - -

_____   _____

3

_____   _____

- - - - - - - -   - - - - - - - -

_____   _____

DIRECTIONS **1–3.** Put ten two-color counters in a shaker. Spill the counters on the page. Sort by color. Beginning with the car behind the engine on the left, color the cars to match the counters of one color, then continue coloring the cars for the other color. Write how many of each color. If one number is greater, circle that number.

Practice

Order Numbers to 10

_____ _____ _____ _____ _____

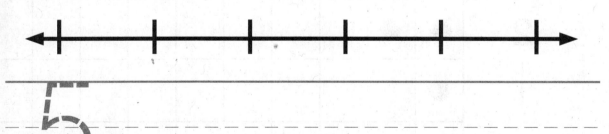

DIRECTIONS 1. How many grapefruits are on each tree? Write the number.
2. Write the numbers in order on the number line.

Practice

Name_____

Number Patterns

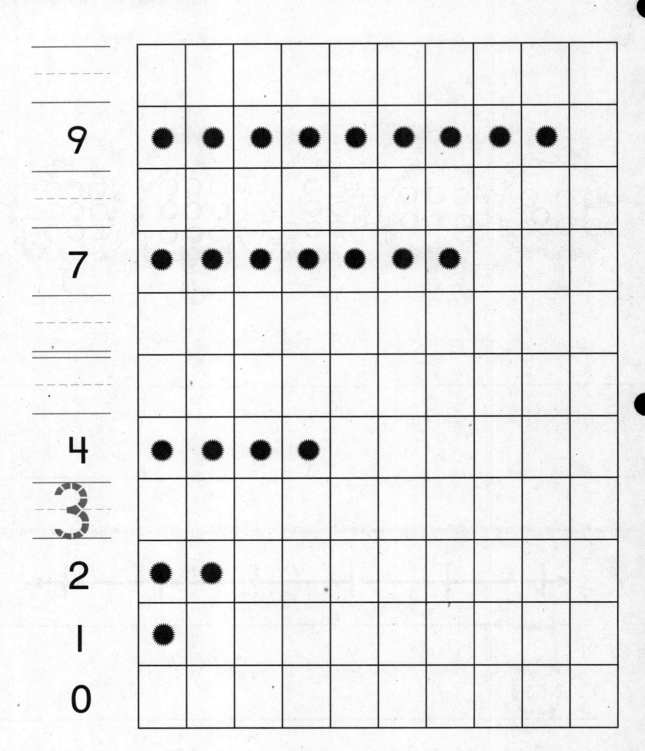

DIRECTIONS Look for a pattern. Tell what you notice about the spots in each row. Draw the spots to follow the pattern. Then write the number to match the pattern.

Practice

Problem Solving Workshop
Skill • Use Estimation

DIRECTIONS Look at the girl at the top of the page. She has ten balloons. Without counting, mark an X on the children that have fewer than ten balloons.

PW33   Practice

Hands On: Make Concrete Graphs

1

Of Which Color Are There Fewest?

2

3

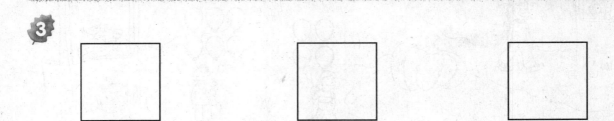

DIRECTIONS **1.** Make a graph with color tiles. Use red to color the first row of tiles. Use blue to color the second row of tiles. Use green to color the third row of tiles. **2.** Use red, blue, and green to color the tiles. Count the tiles. Write how many there are of each color. Circle the least number. **3.** Use red, blue, and green to color the tiles. Circle the kind of tile that there are fewest of in the graph.

Practice

Hands On: Read Concrete Graphs

1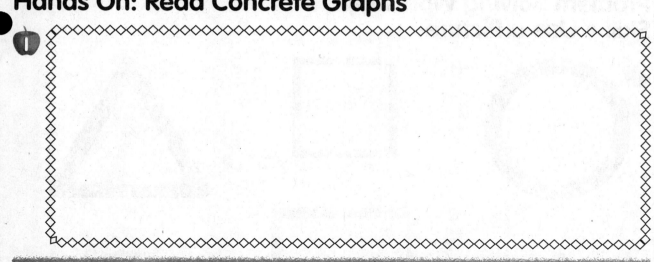

2

How Many of Each Figure?

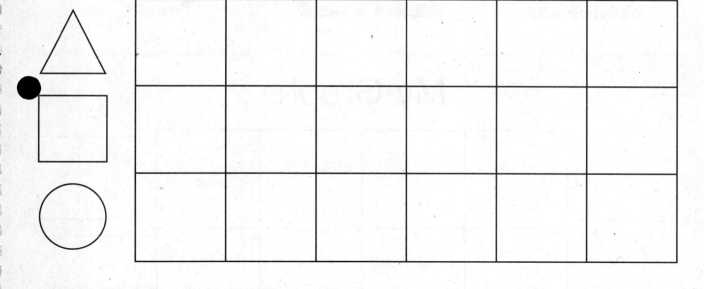

3

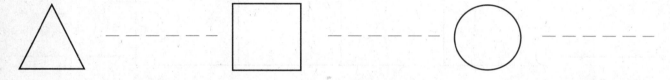

DIRECTIONS 1. Place 6 small blue figures on the workspace. Sort them by shape. 2. Make a graph with your figures. Draw the figures in the graph. 3. Read the graph. Write how many there are of each figure.

Practice

Problem Solving Workshop
Skill • Use a Picture

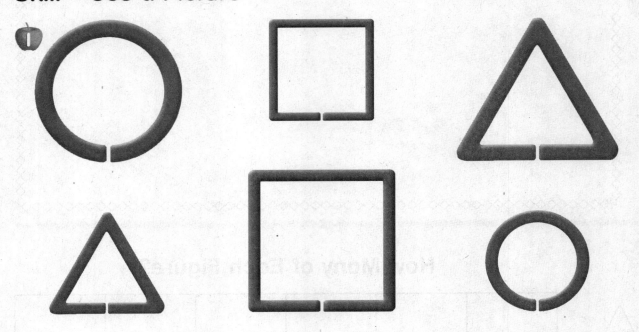

⭐2

My Graph

DIRECTIONS **1. Place figures on the links in the picture. Sort the figures by your own rule. 2. Make a graph with the figures. Write how many there are of each figure.**

Read Picture Graphs

Which Pet Do the Fewest Children Like?

🐹	😊	😊	😊	😊	😊	😊
🐟	😊	😊	😊	😊		
🦜	😊	😊	😊	😊	😊	

 _____  _____  _____

DIRECTIONS 1. Read the graph. Write how many children like each kind of pet.
2. Circle the pet that the fewest children like.

Practice

Make Picture Graphs

🍎 **Which Kind of Animal Are More People Watching?**

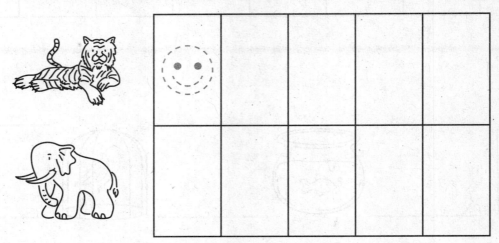

⭐ **2**

_____   _____

_____   _____

DIRECTIONS **1.** Look at the picture. Make a picture graph to show how many people are watching each animal. Circle the row with more people. **2.** Write how many. Circle the greater number.

Problem Solving Workshop
Strategy • Draw a Picture

Spotted and Striped Fish

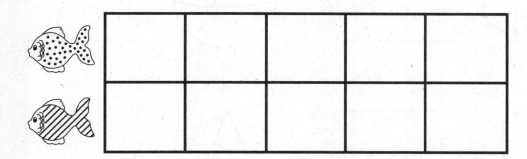

DIRECTION   1. Draw a picture. There are 3 spotted fish. There are 2 more striped fish than spotted fish.   **2.** Show the data on the graph.

Identify and Describe Solid Figures

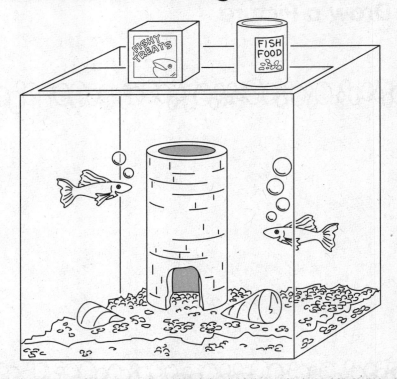

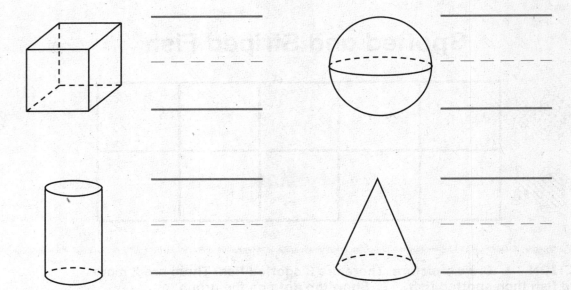

DIRECTIONS **1.** Find solid figures in the picture. Use yellow to circle objects shaped like cubes. Use blue to circle objects shaped like spheres. Use green to circle objects shaped like cylinders. Use red to circle objects shaped like cones.
2. Write how many of each solid figure.

Compare Solid Figures

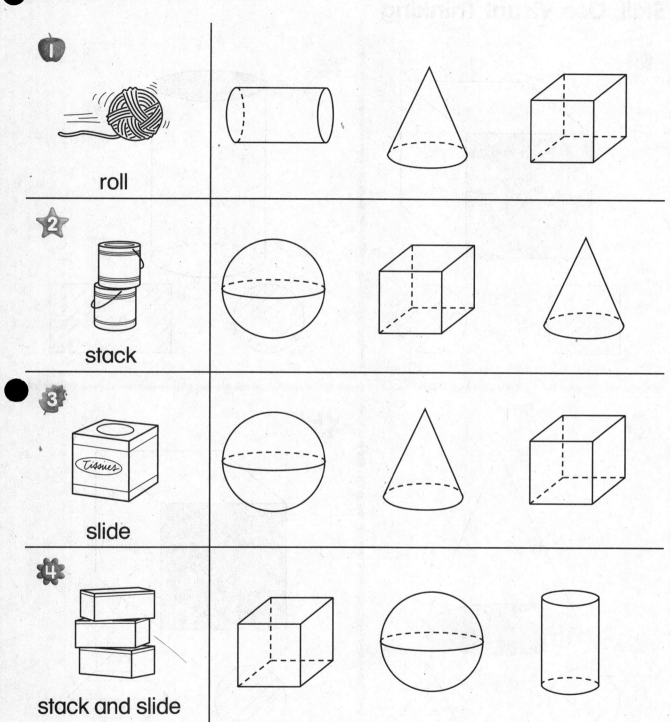

1 roll

2 stack

3 slide

4 stack and slide

DIRECTIONS 1. Mark an X on the figure that does not roll.
2. Mark an X on the figures that do not stack.
3. Mark an X on the figure that does not slide.
4. Mark an X on the figure that does not stack and slide.

Name_____

Problem Solving Workshop
Skill: Use Visual Thinking

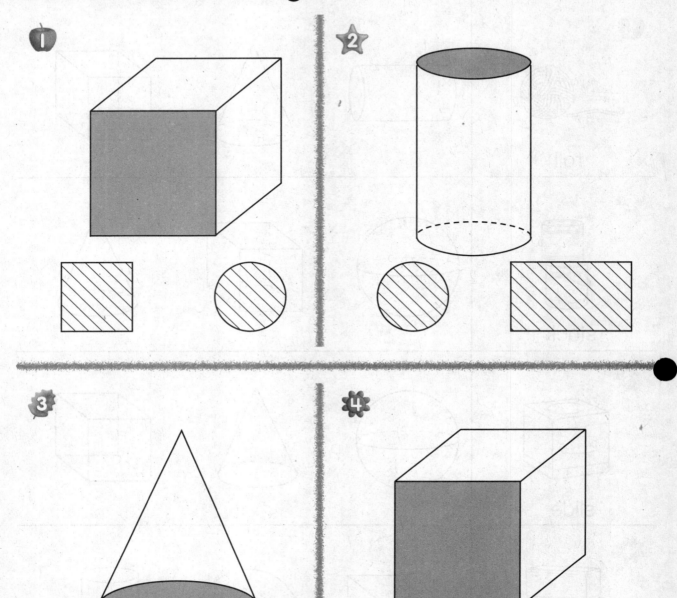

DIRECTIONS 1–4. Circle the plane figure that matches the shape of the shaded surface of the solid figure.

Practice

Lesson 6.4

Hands On: Identify and Describe Plane Figures

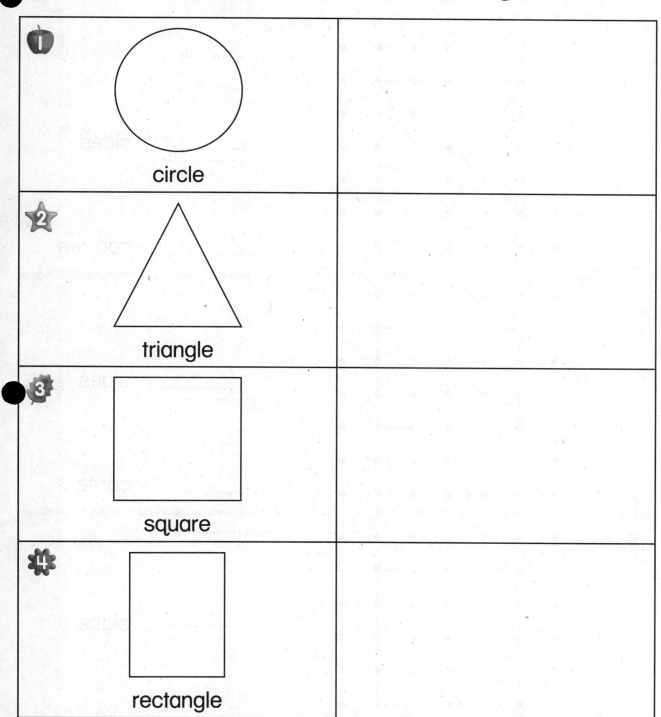

1 circle	
2 triangle	
3 square	
4 rectangle	

DIRECTIONS 1–4. Place a figure that matches the figure. Trace the figure. Describe the figure.

Practice

Compare Plane Figures

1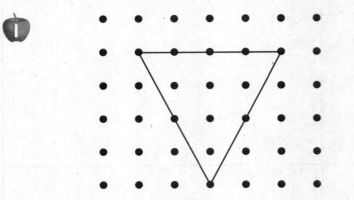

_____

_____ sides

_____ corners

2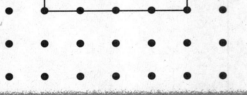

_____

_____ sides

_____ corners

3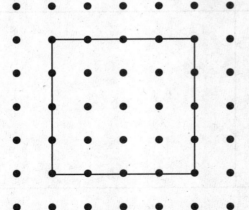

_____

_____ sides

_____ corners

DIRECTIONS 1–3. Trace the outline of the figure. Write how many sides. Write how many corners.

Practice

Problem Solving Workshop
Strategy • Use Logical Reasoning

1

trapezoid   rhombus   triangle   hexagon

2

3

4

5

DIRECTIONS I. Color the trapezoid red, the rhombus blue, the triangle green, and the hexagon yellow. 2–5. Show other ways to use the red, blue, and green figures to cover the outline of the figures. Draw and color the figures.

Hands On: Model, Read, and Write 11, 12, and 13

<table>
<tr><td></td><td></td><td></td><td></td><td></td></tr>
<tr><td></td><td></td><td></td><td></td><td></td></tr>
</table>

<table>
<tr><td></td><td></td><td></td><td></td><td></td></tr>
<tr><td></td><td></td><td></td><td></td><td></td></tr>
</table>

1 12 twelve

2 11 eleven

3 13 thirteen

DIRECTIONS 1–3. Use counters to model the number on the ten frames at the top of the page. Say the numbers as you count. Write the number.

Practice

Read and Write 14, 15, and 16

1 15

2 14

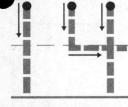

3 16

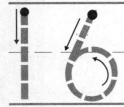

DIRECTIONS 1–3. Count the pets. Say the number as you trace it. Write the number.

Practice

Read and Write 17, 18, and 19

 1

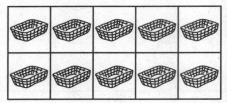 

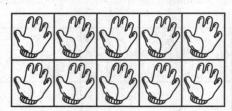 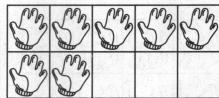

2

19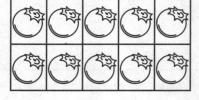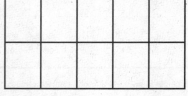

18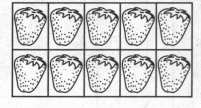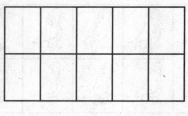

17

DIRECTIONS 1. How many objects? Write the numbers.
2. Draw more fruits to show the numbers. Write the numbers.

Practice

Hands On: Strategy • Use Logical Reasoning

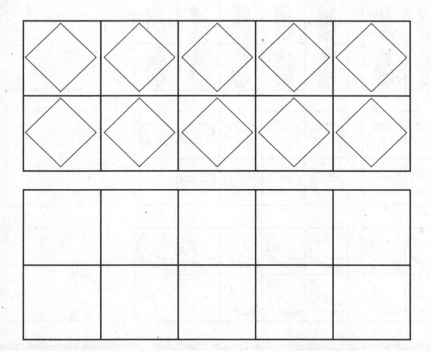

DIRECTIONS 1. Place more tiles on the ten frame to model the number that is 3 more than 17. Write that number. Draw and color the tiles.
2. Place more counters to model the number that is 2 less than 20. Write that number. Draw and color the counters.

Practice

Read and Write 20

 1

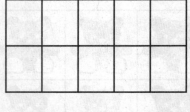

_____

- - - - - -

_____

- - - - - -

_____

- - - - - -

_____

⭐ **2**

18

20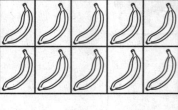

19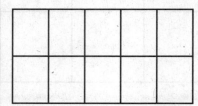

_____

- - - - - -

_____

_____

_____

_____

DIRECTIONS 1. How many pieces of fruit? Write the numbers. 2. Draw more fruit to show the numbers. Write the numbers.

Practice

Count Sets to 20

1

2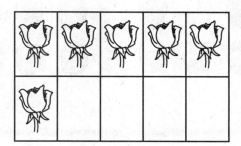

_____

- - - - - -

_____

3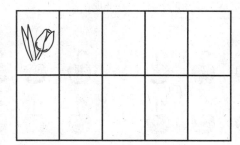

_____

- - - - - -

_____

4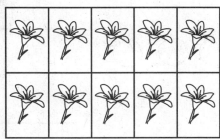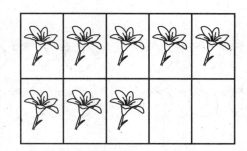

_____

- - - - - -

_____

Practice

Compare Sets to 20

_____

- - - - - - - - -

_____

_____

- - - - - - - - -

_____

_____

- - - - - - - - -

_____

_____

- - - - - - - - -

_____

DIRECTIONS 1–2. Write how many fruit are in each set. Circle the number that is less. Circle the set with fewer fruit.

Order Numbers to 20

1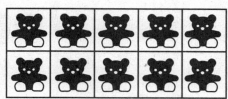

_____

_____

_____

2

- - - - - - - - - -

_____

_____

3

- - - - - - - - - -

_____

_____

4

- - - - - - - - - -

_____

_____

5

___ 16 ___ 18 ___ ___

DIRECTIONS 1–4. How many bear counters? Write the number.
5. Write those numbers in order on the number line.

Skill • Use Estimation

1 🍎

10   20

2 ⭐

10   20

3 🍂

10   20

4 🌼

10   20

DIRECTIONS 1–4. Without counting, circle to show whether each set is closer to 10 pieces of fruit or closer to 20 pieces of fruit.

Practice

Hands On: Model, Read, and Write 21 and 22

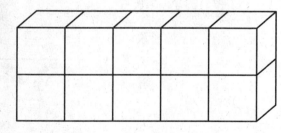

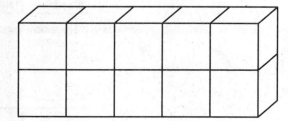

_____

- - - - - - - - - - - -

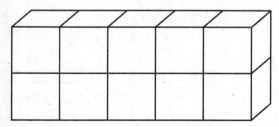

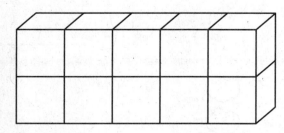

_____

- - - - - - - - - - - -

_____

DIRECTIONS   1. How many more cubes do you need to model 21? Draw the cubes. Write the number in all.   2. How many more cubes do you need to model 22? Draw the cubes. Write the number in all.

Read and Write 23 and 24

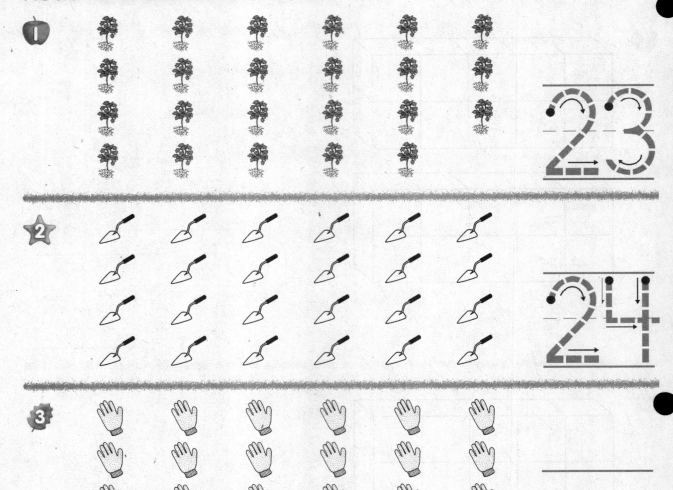

DIRECTIONS 1–2. How many? Trace the number.   3–4. How many? Write the number.

Practice

Problem Solving Workshop
Strategy • Make a Model

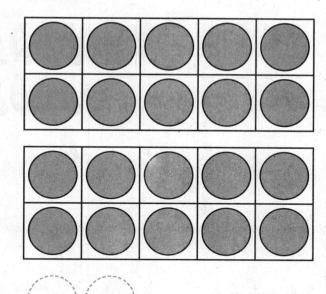

DIRECTIONS 1. Trace the number. Place counters on the outlines to model the number. Trace the counters. 2. Write the number 2 more than 23. Model the number using the ten frames and more counters. Draw and color the counters.

PW57   Practice

Name _____

Lesson 8.4

Read and Write 25

1

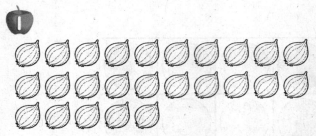

2

3

4

5

6

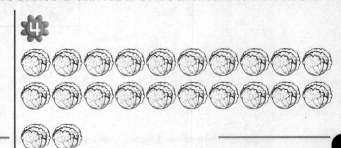

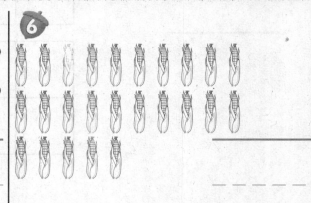

DIRECTIONS   **1.** How many vegetables? Trace the number.   **2–6.** How many vegetables? Write the number.

Practice

Name_____

Read and Write 26 and 27

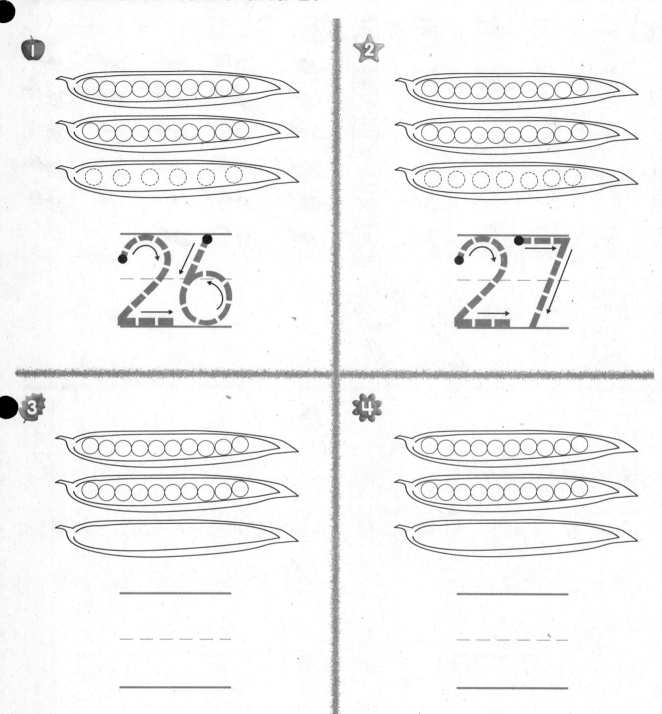

DIRECTIONS 1–2. Trace the peas. How many? Trace the number in all. 3. How many peas? How many more peas do you need to make 26? Draw the peas. Write the number in all. 4. How many peas? How many more peas do you need to make 27? Draw the peas. Write the number in all.

Practice

Name_____

Read and Write 28 and 29

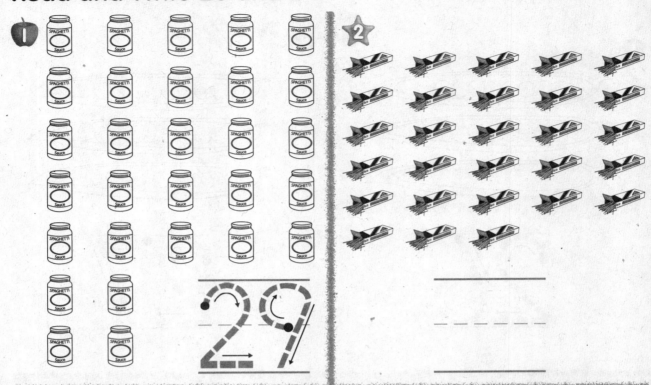

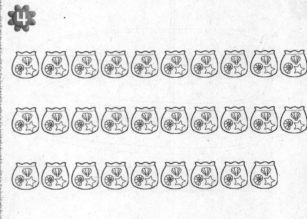

DIRECTIONS   1. How many jars of pasta sauce? Trace the number. 2–4. How many bags of pasta? Write the number.

Practice

Read and Write 30

 1

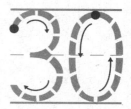

 2

 3

4

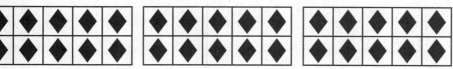

DIRECTIONS 1. How many shapes? Trace the number. **2–4.** How many shapes? Write the number.

Problem Solving Skill • Use Data from a Picture

_____

- - - - - - - - - - -

_____

- - - - - - - - - - -

_____

_____

- - - - - - - - - - -

_____

- - - - - - - - - - -

_____

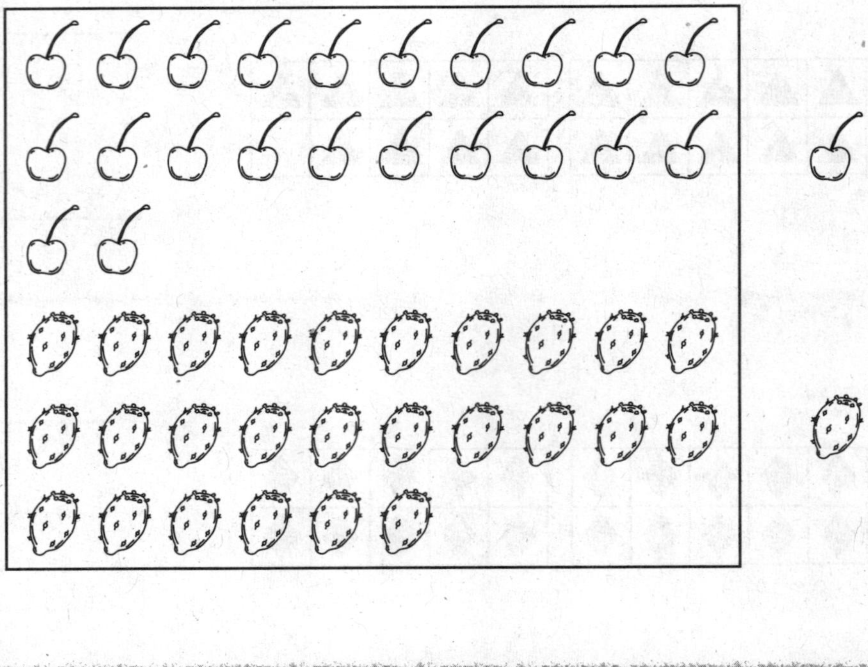

DIRECTIONS I. Write how many cherries are in the picture. Write how many pears are in the picture. Circle the number that shows more fruit. **2.** Circle the fruit of which one picture shows 2 fewer pieces than the other picture.

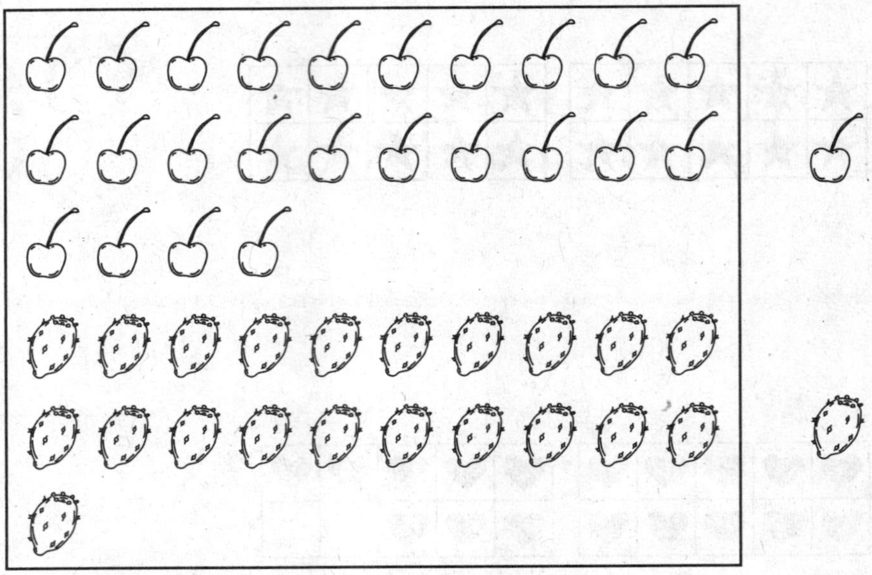

Practice

Order Numbers to 30

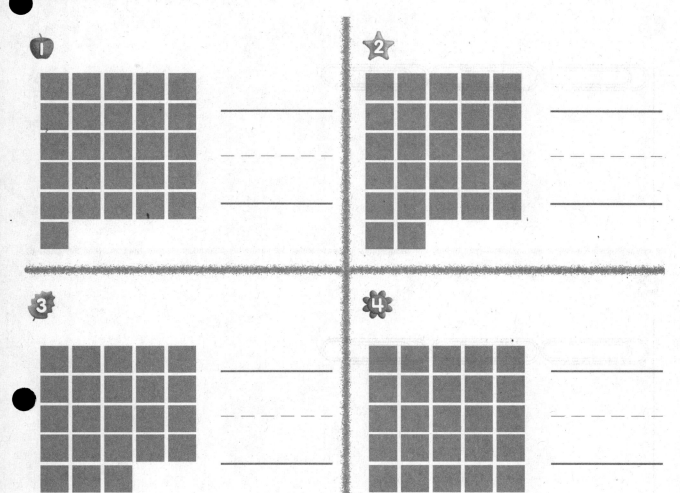

① _____

_____

_____

② _____

_____

_____

③ _____

_____

_____

④ _____

_____

_____

⑤

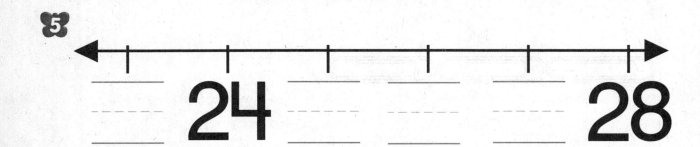

24 ___ ___ ___ 28

DIRECTIONS 1–4. How many color tiles? Write the number.
5. Write those numbers in order on the number line.

Compare Length

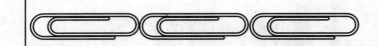

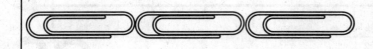

DIRECTIONS 1. Make a paper clip chain that is longer. Draw the paper clip chain. 2. Make a paper clip chain that is the same length. Draw the paper clip chain. 3. Make a paper clip chain that is shorter. Draw the paper clip chain.

Practice

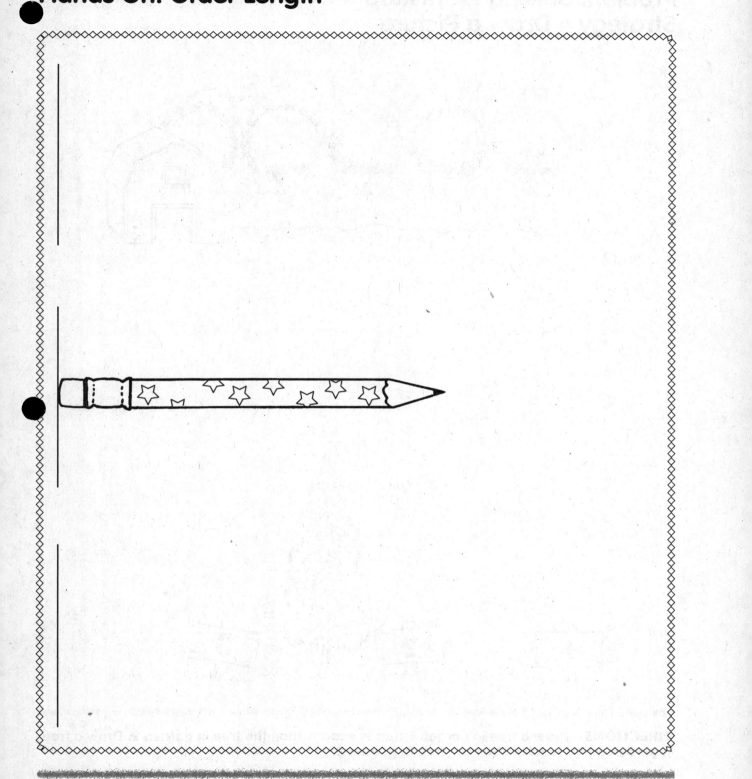

Hands On: Order Length

DIRECTIONS   Find a classroom object that is shorter than the pencil and an object that is longer than the pencil. Draw the objects in order from shortest to longest.

Problem Solving Workshop
Strategy • Draw a Picture

1

DIRECTIONS Draw a tree in garden 1 that is shorter than the tree in garden 2. Draw a tree in garden 3 that is taller than the tree in garden 2.

Hands On: Explore Capacity

about _____
_____

about _____
_____

about _____
_____

about _____
_____

DIRECTIONS 1–4. Use containers like the ones pictured. Fill each container with scoops of dried beans. Write about how many scoops each container holds.

Hands On: Compare Capacity

holds more

holds the same

holds less

holds more

holds the same

holds less

---

holds more

holds the same

holds less

DIRECTIONS   I. Use dried beans to fill a food storage container like the one pictured. **2–4.** Pour the beans from the container you filled in ① into a storage container like the one pictured. Does this container hold more, less, or the same amount of beans as the first container? Circle your answer.

Practice

Problem Solving Workshop
Strategy • Use a Picture

DIRECTIONS Look for similar objects in different sizes. Use green to color the containers that hold the most. Use orange to color the containers that hold the least.

Practice

Hands On: Explore Weight

Left ✋   **Right** ✋

DIRECTIONS   1–4. Find the first object in the row, and hold it in your left hand. Find the rest of the objects in the row, and take turns holding each object in your right hand. Circle the object that is lighter than the object in your left hand.

Hands On: Compare Weight

DIRECTIONS Find three classroom objects that have different weights. Draw the objects in order from lightest to heaviest.

Days of the Week

April						
Sunday	Monday	Tuesday	Wednesday	Thursday	Friday	Saturday
			1	2	3	4
5	6	7	8	9	10	11
12	13	14	15	16	17	18
19	20	21	22	23	24	25
26	27	28	29	30		

2
Tuesdays 4 _____

3
Saturdays _____
_____

4
Wednesdays _____
_____

5
Days in April _____
_____

DIRECTIONS 1. Circle all the Saturdays. Draw an X on all the Thursdays. 2. Write how many Tuesdays are in this month. 3. Write how many Saturdays are in this month. 4. Write how many Wednesdays are in this month. 5. Write how many days are in April.

Explore Sequence of Events

yesterday	today	tomorrow

SUNDAY	MONDAY	TUESDAY	WEDNESDAY	THURSDAY	FRIDAY	SATURDAY

_____ _____ _____

_____ _____ _____

morning	afternoon	evening

DIRECTIONS **I.** Draw a line from *today* to the name of the day. Point to the name of the day before. Draw a line from that day to *yesterday*. Point to the name of the day after today. Draw a line to *tomorrow*. **2.** Look at each picture. Draw a line to the time of day this would probably happen. Use numbers to show the order.

Name_____

Explore a Calendar

August						
Sunday	Monday	Tuesday	Wednesday	Thursday	Friday	Saturday
						1
2	3		5			8
9	10	11			14	15
16	17		19	20	21	22
23	24	25		27		29
30	31					

DIRECTIONS. Trace the name of the month at the top. Use red to color the names of the days of the week. Trace the numbers and write the missing numbers. Circle the first and last days of the month.

Practice

Problem Solving Workshop
Skill • Use a Calendar

January	February	March
April	May	June
July	August	September
October	November	December

July						
Sunday	Monday	Tuesday	Wednesday	Thursday	Friday	Saturday
			1	2	3	4
5	6	7	8	9	10	11
12	13	14	15	16	17	18
19	20	21	22	23	24	25
26	27	28	29	30	31	

DIRECTIONS 1. Use red to color the first month of the year. Use yellow to color the last month of the year. Circle the month that comes right after June. 2. Use blue to color the day that comes right after Saturday. Circle the day that is the fourth of this month. Use yellow to color all the Mondays in this month.

Practice

Name_____

More Time, Less Time

1

2

3

4

DIRECTIONS 1–2. Circle the activity that usually takes more time. 3–4. Circle the activity that usually takes less time.

Practice

Use a Clock

1

about

6

o'clock

2

about

_____

_____

o'clock

3

about

_____

_____

o'clock

4

about

_____

_____

o'clock

DIRECTIONS   1–4. The minute hand is missing! Use just the hour hand. About what time does the clock show? Write your answer.

Time to the Hour

 1

 2

 3

 4

DIRECTIONS **1–2.** Look at the picture. Draw the hour hand to show about what time it might be. **3–4.** Look at the picture. Write a number to show about what time it might be.

Practice

Name_____

Problem Solving Workshop
Strategy • Draw a Picture

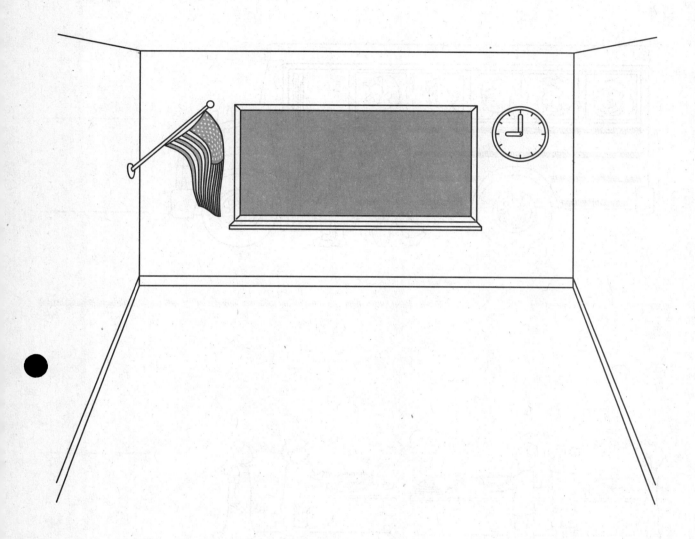

DIRECTIONS Draw a picture to show something you do at 9 o'clock in the morning.

PW79   Practice

Problem Solving Workshop
Strategy • Act It Out

_____

- - - - - - - - -

_____

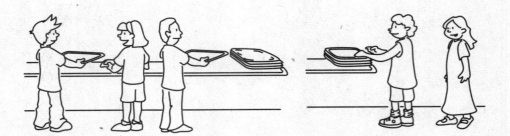

_____

- - - - - - - - -

_____

DIRECTIONS 1–2. Tell an addition story about the picture. Act out the story. Write the number that shows how many children there are in all.

Practice

Hands On: Model Addition

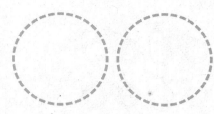

3   2

---

4   3

---

7   2

DIRECTIONS 1–3. Tell a story using the numbers. Model the story with counters. Draw the counters. Write the number that shows how many in all.

Name _____

Joining Groups

_____   and   _____   is   _____

_____   and   _____   is   _____

_____   and   _____   is   _____

DIRECTIONS 1–3. Write how many in each group. Circle the two groups. Write how many in all.

Practice

Introduce Symbols to Add

1

2	and	3	is	5

2

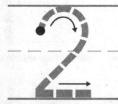 

4	and	6	is	10

3

3	and	3	is	6

DIRECTIONS 1–3. Write how many in each group. Circle the two groups. Trace the symbols. Write how many in all.

Addition Patterns

$$3 + 1 = $$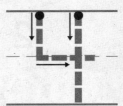

$$4 + 1 = $$

$$5 + 1 = $$

$$6 + 1 = $$

DIRECTIONS 1–4. How many leaves? Draw one more leaf. Write the number of leaves in all to complete the addition sentence.

Practice

Name_____

 Lesson 11.6

Addition Sentences

_____ + _____ ===

_____ + _____ ===

 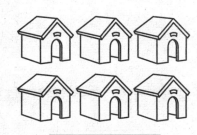

_____ + _____ === _____

 DIRECTIONS 1–3. Tell a story about the objects. Complete the addition sentence.

Practice

Hands On: Create and Model Addition Problems

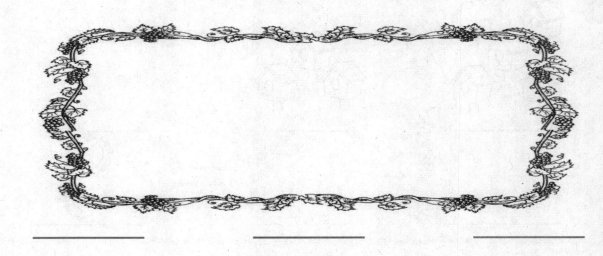

_____   +   _____   =   _____

_____   _____   _____

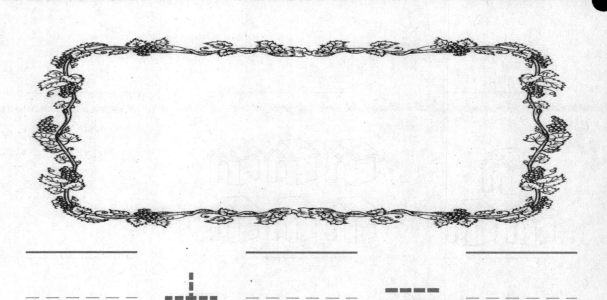

_____   +   _____   =   _____

_____   _____   _____

DIRECTIONS 1–2. Tell an addition story. Model your story with cubes.
Draw the cubes. Complete the addition sentence.

Name_____

Problem Solving Workshop
Skill • Use a Model

_____   🐻 + _____ 🐻 === 10

_____

_____   🐻 + _____ 🐻 === 10

_____

3

_____   🐻 + _____ 🐻 === 10

_____

DIRECTIONS   1–3. Use two colors of bear counters to show different ways to make 10. Color the counters. Complete the addition sentence for each model.

Practice

Name _____

Problem Solving Workshop
Strategy: Act It Out

 1

_____

_ _ _ _ _ _ _ _ _ _ _ _ _ _

_____

 2

_____

_ _ _ _ _ _ _ _ _ _ _ _ _ _

_____

DIRECTIONS 1–2. Tell a subtraction story about the picture. Act out the story. Write the number that tells how many children are left.

PW88   **Practice**

Hands On: Model Subtraction

9   6   _____
              - - - - - -

              _____

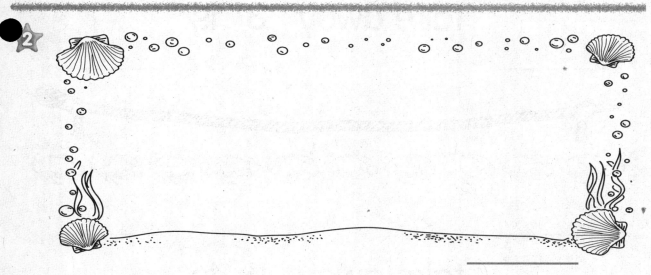

6   4   _____
              - - - - - -

              _____

DIRECTIONS 1. *Nine fish swam in a pond. Six fish swam away. How many fish are left?* Model the story with cubes. Write the number that tells how many are left. 2. *Six turtles crawled along the beach. Four turtles walked into the sea. How many turtles are left?* Model the story with cubes. Write the number that tells how many are left.

Name_____

Separating Groups

_ _ _ _ _ _ _ _ _   take away   3   is   _ _ _ _ _ _ _

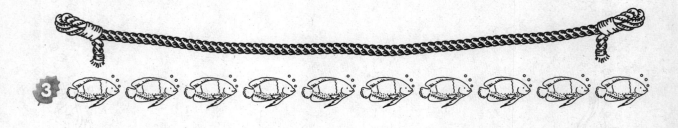

_ _ _ _ _ _ _ _ _   take away   3   is   _ _ _ _ _ _ _

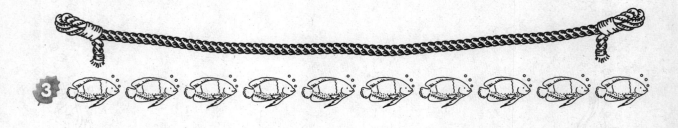

_ _ _ _ _ _ _ _ _   take away   4   is   _ _ _ _ _ _ _

DIRECTION   1–3. Write how many there are in all. Mark an X on the animals that are taken away. Write how many are left.

Use Symbols to Subtract

1

7   take away   5   is   2

_____   _____   _____

------   =====

_____   _____

2

9   take away   5   is   4

_____   _____   _____

------   =====

_____   _____

3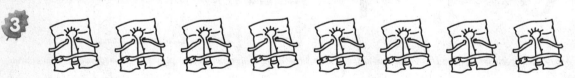

8   take away   3   is   5

_____   _____   _____

------   =====

_____   _____

DIRECTIONS   1–3. Write how many objects there are in all. Mark an X on the objects that are taken away. Complete the subtraction sentence to show how many are left.

Subtraction Patterns

1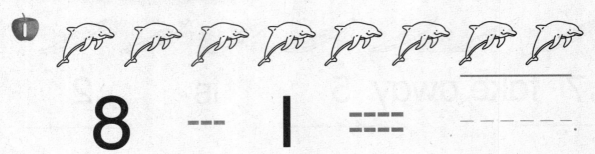

$$8 - 1 = \underline{\hspace{2cm}}$$

2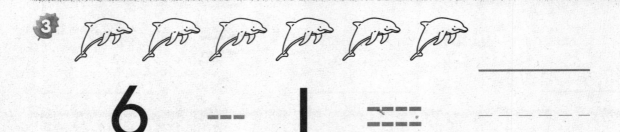

$$7 - 1 = \underline{\hspace{2cm}}$$

3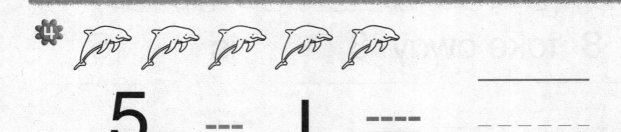

$$6 - 1 = \underline{\hspace{2cm}}$$

4

$$5 - 1 = \underline{\hspace{2cm}}$$

DIRECTIONS   1–4. How many dolphins are there in all? Mark an X on the dolphin that is taken away. Complete the subtraction sentence to show how many dolphins are left.

Practice

Subtraction Sentences

1

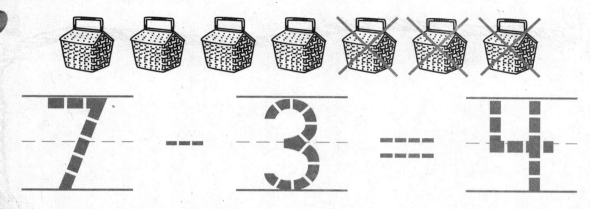

2

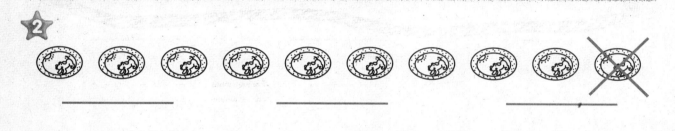

3

DIRECTIONS 1–3. Tell a story about the picnic items. Complete the subtraction sentence.

Practice

Hands On: Create and Model Subtraction Problems

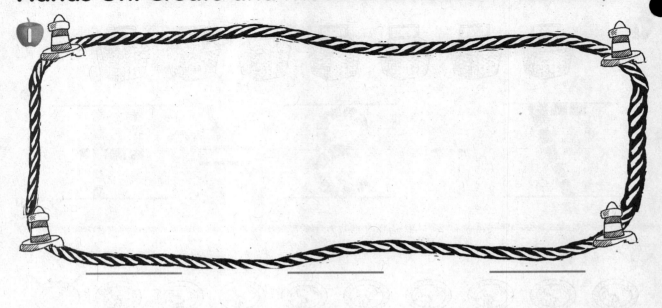

DIRECTIONS 1–2. Tell a subtraction story. Model your story with cubes. Draw the cubes. Mark an X on the cubes that are taken away. Complete the subtraction sentence.

Practice

Problem Solving Workshop
Skill: Use a Model

10 --- _____ === _____
    _____   _____

10 --- _____ === _____
    _____   _____

10 --- _____ === _____
    _____   _____

10 --- _____ === _____
    _____   _____

DIRECTIONS 1–4. Use counters to show different ways to subtract from 10.
Mark an X on the counters you take away. Complete the subtraction sentence
for your model.

Practice